# What Makes a Plant a Plant?

by Gary Rushworth

## Table of Contents

# Introduction

**P**lants have been on Earth since before the dinosaurs. Plants have been here for hundreds of millions of years. Did you know that some plants living today are over four thousand years old? Do you know how many kinds of plants there are? More than you might think!

In this book, you will discover what makes a plant a plant. You'll find out how plants are different from other living things. You'll also begin to look at plants the way scientists do.

The Methuselah bristlecone pine in California is over 4,700 years old. ▶

◀ Plants were on Earth long before the dinosaurs.

More than 400,000 types of plants grow on Earth. It is hard to keep track of them all. Scientists have grouped all plants into a kingdom. It is called the Kingdom Plantae, or plant kingdom.

In this book, you will learn about the plant kingdom. You will learn about different kinds of plants. You'll discover what plants need to survive and protect themselves. You will learn how plants help the environment. Read on and learn about what makes a plant a plant!

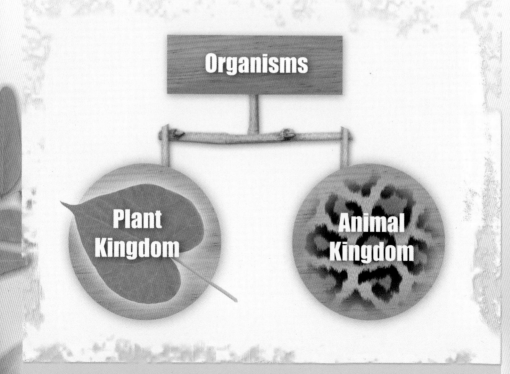

▲ Plants and animals are classified into different kingdoms.

# What Makes a Plant a Plant?

There are millions of types of living things on Earth. Scientists have devised a system of **classification** to help keep track of all the types of **organisms**. Scientists place organisms into groups. Organisms that share the same traits are grouped together.

The first system of classification divided organisms into two kingdoms: the plant kingdom and the animal kingdom. As scientists learned more about organisms and discovered new living things, they added new kingdoms. Today there are five kingdoms.

**Biological Building Blocks**

The cell is the basic building block of all living things. Plants, animals, and humans are all made of cells.

Plant cells and animal cells contain many of the same parts. These parts act in similar ways. Plant and animals cells are different in some ways, too.

All cells are about sixty percent water. The water and carbon compounds make up the semi-fluid portion of a cell, the cytoplasm. The parts of a cell are called organelles. Organelles help a cell make energy, store food, eliminate wastes, reproduce, and communicate with other cells. The organelles float in the cytoplasm within the **cell membrane**. The cell membrane acts like a container for the cytoplasm and organelles. All cells have a cell membrane.

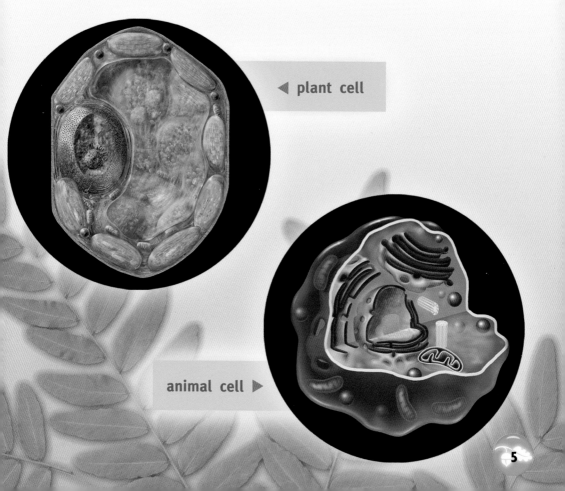

◀ plant cell

animal cell ▶

## The Plant Kingdom

Each plant cell has a cell wall. An animal cell does not have a cell wall. The cell walls give a plant shape. The cell walls also hold a plant upright. This is important because plants need to withstand pressure from wind, rain, and snow.

The nucleus is an organelle near the center of a cell. The nucleus carries all the information needed to make a new cell. The nucleus tells a cell when and how to get food, when to make energy, and when to make a new plant.

cell wall

▲ Plant cells have a cell wall.

Plant cells have special parts that are not found in the cells of any other organism. The special organelles are called chloroplasts (KLOR-uh-plasts). Chloroplasts contain **chlorophyll** (KLOR-uh-fil). Chlorophyll is the chemical that gives a plant its green color.

Chlorophyll reacts with light to allow a plant to make its own food. Plants make food through a complex series of events. Together, these events are called **photosynthesis** (foh-toh-SIN-theh-sis). Photosynthesis means "to mix with light."

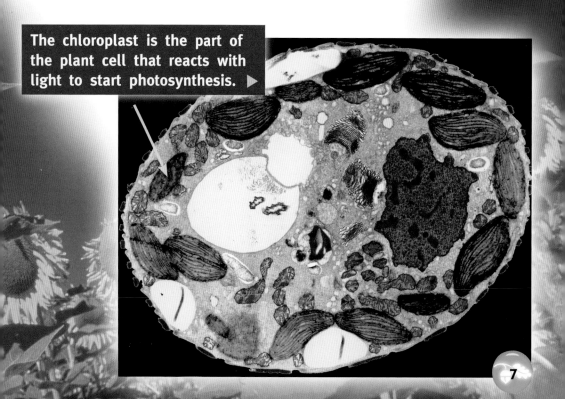

The chloroplast is the part of the plant cell that reacts with light to start photosynthesis. ▶

**How Photosynthesis Works**

Plants absorb a gas called carbon dioxide through tiny holes in their leaves. They take in water through roots in the soil. Sunlight causes chlorophyll to react with the water and carbon dioxide. The result is a kind of sugar that the plant uses for food.

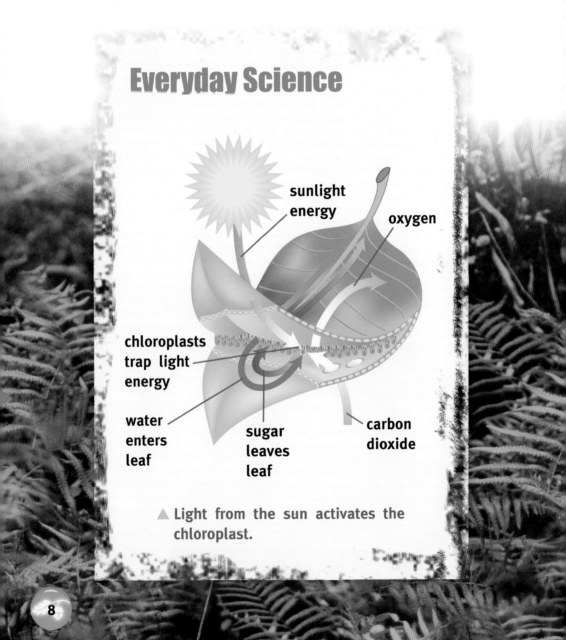

**Everyday Science**

sunlight energy

oxygen

chloroplasts trap light energy

water enters leaf

sugar leaves leaf

carbon dioxide

▲ Light from the sun activates the chloroplast.

In the process of using water and carbon dioxide to make sugar, plants release oxygen into the air as a waste product. The air we breathe in contains enough oxygen to support life. When we exhale, we breathe out carbon dioxide. The same is true for animals.

People and animals help plants by supplying them with carbon dioxide. Plants help animals and people by replacing the oxygen in the air we breathe.

✔ Point

**Think About It**

How else do plants help people? Make a list. Put a check mark beside each idea you read about in this book.

# The Plant Kingdom

**S**cientists use classification systems to record all the plants known today. Classification systems are based on natural characteristics, or similarities between individual plants. The science of classification is known as taxonomy.

Plants are first grouped according to how they get water. **Vascular** plants get water through tubes in their stems. Nonvascular plants bring water into their stems and leaves through tiny holes.

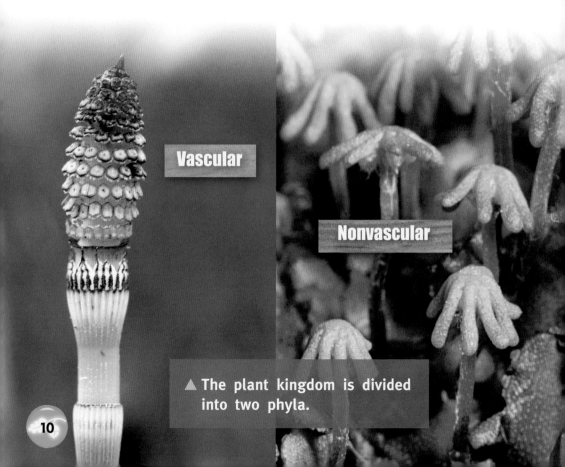

Vascular

Nonvascular

▲ The plant kingdom is divided into two phyla.

Plants are next divided by whether they produce seeds. Then they are divided by whether they live on land or in water, by their root systems, and so on. Each grouping defines a plant in detail.

There are seven classification groups in all. Each time a grouping is done, the number of differences between the members becomes smaller. The kingdom is the largest group. The smallest grouping is the **species**.

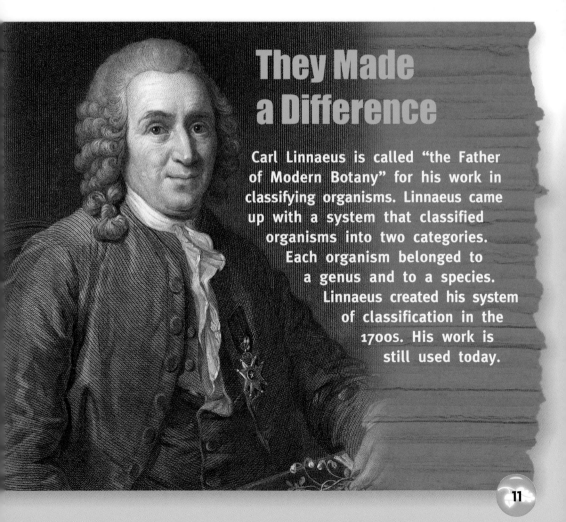

# They Made a Difference

Carl Linnaeus is called "the Father of Modern Botany" for his work in classifying organisms. Linnaeus came up with a system that classified organisms into two categories. Each organism belonged to a genus and to a species. Linnaeus created his system of classification in the 1700s. His work is still used today.

## Seed Plants

Seed plants include trees and flowering plants. Each unique group of plants is called a species. For example, roses are one species. Tulips are another species. Both species are plants and both species have flowers.

## Nonseed Plants

Nonseed plants grow from **spores**. Spores are one-celled structures that grow into new plants. Some examples of nonseed plants that make new plants using spores include mosses, liverworts, and ferns.

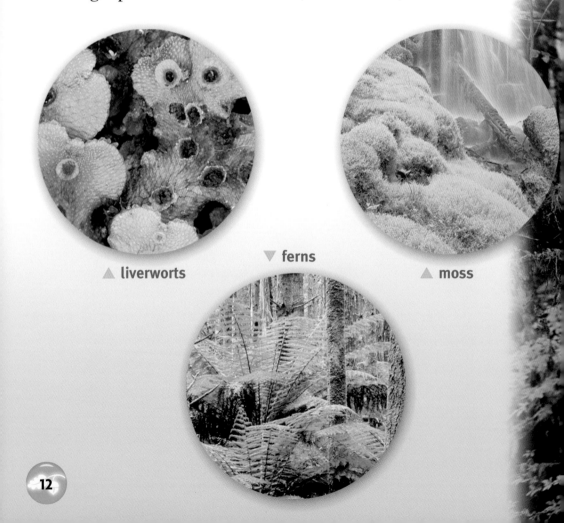

▲ liverworts

▼ ferns

▲ moss

## Liverworts and Mosses

Both mosses and liverworts grow in moist places. Because they are nonvascular plants, they have no way of storing water in their system. They live close to the ground on rocks, soil, and even on other plants. They never have flowers.

**Everyday Science**

Mosses and liverworts may be small, but they play a big part in our environment. They help prevent erosion and act as a foundation for other plants to grow upon.

**Ferns**

Like mosses, ferns grow from spores instead of seeds. They are different because they carry water through tubes. The leaves of ferns are called fronds. The spores of a fern are found on the underside of the fronds.

Ferns are one of the oldest plants we know of. Scientists believe that ferns were on Earth more than 300 million years ago! Today scientists have classified more than 11,000 species of ferns. Most grow in tropical areas.

## Careers Botanist

A botanist is someone who studies plants. Over 500,000 kinds of plants are classified already, and more are being discovered every day!

## Horsetails

Horsetails are another kind of plant that have no seeds. Today scientists know of about forty different species of horsetails. They are usually found in swamps or marshy areas.

### Everyday Science

Fossil evidence indicates that many extinct horsetails were treelike and attained a far greater size than the ones we see today.

◄ spores on fern leaves

**Plants with Seeds**

There are two kinds of plants that reproduce through seeds. They are conifers and flowering plants. A conifer carries its seeds in a cone, like a pinecone. Scientists know of about 600 species of conifers. Pine trees, fir trees, spruces, cedars, and yew trees are all conifers. One way you can tell if a tree is a conifer is to look at its leaves. Conifers usually have leaves that look like needles.

Can you think of an example of a flowering plant? It's not hard—there are more than 200,000 species of flowering plants! A rosebush and a cherry tree are two species of flowering plants.

◀ cherry trees

◀ pine tree

▲ cedar tree needles

## Monocots and Dicots

Flowering plants are further classified as monocots or dicots. Both types have roots, stems, and leaves. They both produce new plants from seeds. The difference is in the seeds. The seed stores food for the developing new plant. This food source is called the cotyledon (kah-tih-LEE-dun). The cotyledon looks like a tiny leaf. Monocots produce one cotyledon. Orchids, lilies, and irises are all monocots.

crane flowers ▷

Dicots produce two cotyledons. A banana tree and a mint plant are both dicots. One way to tell the difference between a monocot and a dicot is to look at their leaves. In a monocot, the leaf veins are parallel—they run next to each other. In a dicot, the leaf veins grow in a branch pattern.

toad lily ▼

# Everyday Science

## Classification of Plants

| | |
|---|---|
| Kingdom | Separates plants from animals |
| Phylum | Separates plants as vascular or nonvascular |
| Class | Separates plants into flowering and nonflowering varieties |
| Order | Separation of plants by common ancestors |
| Family | Group of plants with many similar characteristics |
| Genus | Group of plants containing two or more species |
| Species | Level that defines the individual plant |

English ivy ▶

## Making New Plants

In order for a species to survive, the organisms within a species must reproduce. A plant is no exception. Sometime during its life cycle, a plant must make another plant.

Some plants begin as seeds or spores. Other plants simply grow from part of a parent plant. Seeds, spores, and individual plant cells contain all the material needed to make a new cell or plant. We call this genetic material.

Plants reproduce in two ways. Some plants have male and female parts. These parts produce female ova that must be fertilized by male sperm. This is called sexual reproduction. Some plants reproduce by making new cells from existing cells. This is called **vegetative reproduction**.

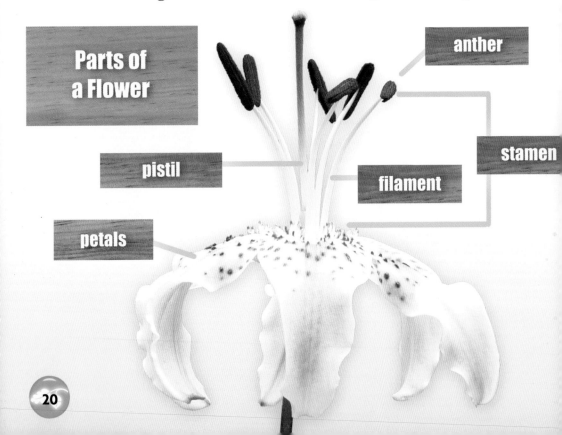

Parts of a Flower

anther

stamen

pistil

filament

petals

In order for sexual reproduction, or fertilization, to take place, plants need to spread their pollen to other plants. This process is **pollination** (pah-lih-NAY-shun). The pollen contains the male reproductive cells, or sperm, of one plant. Those cells come in contact with the ova, or eggs, of another plant. When a male sperm combines with a female ovum, the ovum becomes fertilized, or ready to make a new plant.

Once fertilized, seeds and spores separate from the plant. Some fall to the ground near the parent plant. Some travel long distances through the air or on the water. The seeds germinate (JER-mih-nate), or start to grow. The cycle continues, adding generation after generation of new plants.

## Everyday Science

Seeds and spores are spread by the wind, water, and animals. When seeds or spores land, they need the right conditions to grow. Sunlight, water, soil, and proper temperatures provide the right conditions.

# What Do Plants Need to Survive?

We know that some plants, like ferns, have been on Earth for millions of years. Other plants that once existed are now extinct. Plants become extinct as their habitats change. For example, if the temperature in a region changes, some plants may not be able to adapt to the new environment. Sometimes a species will die out because another species will deprive it of the light it needs to survive.

## Math Matters

It is estimated that nearly 1,000 species of vascular plants have become extinct in the past 100 years. Many more are considered endangered. These are just a few.

- American hart's tongue fern
- dwarf lake iris
- eastern prairie fringed orchid
- Fassett's locoweed
- Houghton's goldenrod
- Michigan monkey flower
- northern wild monkshood
- Pitcher's thistle
- prairie bush-clover

iris

daisy

Plant extinction is hard to prove. Plants can exist for years as dormant, or sleeping, seeds. When conditions are right, they may grow again. Plants once believed to be extinct have turned up years later in isolated locations.

Plants have also developed ways to protect themselves. Some plants are poisonous, so animals and insects don't eat them. Other plants have sharp thorns that help keep hungry animals away.

Some plants, like rosebushes, have sharp thorns to protect them. ▶

thorns

orchid

## Other Dangers

Bad weather, high winds, heavy rain, floods, and drought can damage and destroy plants. Plants have adaptations to these conditions. Some plants make seeds that are waterproof. Other plants have extensive root systems to search for water. Some plants use high heat from the sun or a fire to release their seeds. Some plants can grow in snow and cold or underwater.

▼ The wide root systems of trees help them find water needed for survival.

### ✔ Point
Picture It

Design a new plant. Give it features that protect it from weather, animals, insects, and humans. Name your plant. Then write a caption for your picture.

Humans are among the biggest threats to plant survival. A careless person who drops a match in a forest can start a fire that destroys thousands of plants. As human population grows, more land is cleared to build cities, roads, and houses. Many trees and other plants are destroyed.

Humans have relied on plants for thousands of years. Plants provide food. People use trees to build houses. Plants also provide fibers that people use to make clothing. Many medicines are made from plants.

## Careers

### Herbalist

An herbalist is someone who heals people using only natural medicines made from plants. Some herbalists treat humans and animals.

**horehound herb**

Horehound is named for Horus, the Egyptian god of sky and light. Horehound was used to treat fevers and snakebites.

**comfrey**

Comfrey helps speed up the natural replacement of body cells. People apply comfrey to promote healing in damaged or injured tissues.

**red rosehip**

Rosehip is high in vitamin C and is often used to make tea.

**ginger**

Ginger root is used to treat nausea caused by motion sickness and other illnesses.

**chives**

The Chinese used chives medicinally as long as 3,000 years ago. The Romans believed chives could relieve the pain from sunburn or a sore throat.

## Science and Plants

Newer classification systems using genetic connections help to fill in the blanks about the origins of plants, their habitats, and how they respond and adapt to changes in the environment.

Genetic systems of classification are being used more and more. Genetic classification systems use a plant's deoxyribonucleic acid, or DNA. Genetic information helps us to trace the way a species of plant has developed over time.

## They Made a Difference

Rosalind Franklin's research was central to the Nobel prize–winning discovery of DNA's double-helix structure. She worked with James Watson and Francis Crick, but she never received the credit she was due during her lifetime.

All organisms have specific DNA patterns. DNA is found in the nucleus of the cell. DNA contains all the information about a plant and its ancestors. The genetic information is passed from one generation to the next.

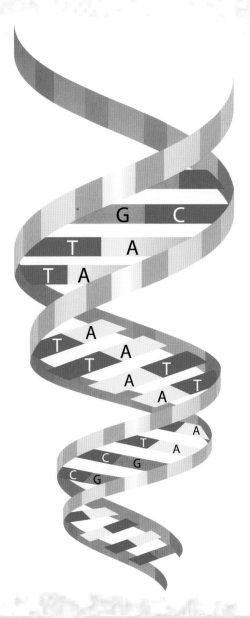

## Everyday Science

The DNA molecule is a long chain of genetic material that looks like a ladder. The two strands of the double helix are anti-parallel, which means that they run in opposite directions.

The sugar-phosphate backbones are on the outside of the helix, and the bases are on the inside. The backbones are like the beams of a ladder. The bases in the middle form the rungs of the ladder. Each rung is composed of two base pairs.

# Conclusion

We see plants or things made from plants just about everywhere. What did you have to eat today? Part of it was probably from a plant.

Plants have been on Earth for billions of years. There are billions of plants on Earth. Scientists believe there are many more waiting to be discovered. Many plants have already become extinct. Many others are endangered.

We depend on plants for our survival. Plants, animals, and people live together in a delicate balance. As long as plants continue to thrive, people will thrive as well.

# Glossary

cell membrane
(SEL MEM-brane) the layer of a cell that holds the cell fluid and organelles in place (page 5)

chlorophyll
(KLOR-uh-fil) the chemical found in the chloroplasts of plant cells that is necessary for photosynthesis (page 7)

classification
(kla-sih-fih-KAY-shun) a system of arranging things into groups (page 4)

organism
(OR-guh-nih-zum) a living thing (page 4)

photosynthesis
(foh-toh-SIN-theh-sis) the process that plants go through to make food (page 7)

pollination
(pah-lih-NAY-shun) the process by which pollen from a male plant part is transferred to a female plant part (page 21)

species
(SPEE-sheez) a category of classification (page 11)

spore
(SPOR) a single cell that produces a new living plant (page 12)

vascular
(VAS-kyuh-ler) a plant that carries water through a system of tubes (page 10)

vegetative reproduction
(veh-jeh-TAY-tiv ree-pruh-DUK-shun) a form of asexual reproduction that takes place in plants (page 20)

# Index